P R E

BY THE SAME AUTHOR

Making Babies
Infertility – a Sympathetic Approach
Getting Pregnant

Genetic Manipulation

ROBERT WINSTON

PHOENIX

Acknowledgements

I am very grateful to two close colleagues, both excellent scientists, for the stimulus they provide and their intellectual support. The encouragement with this manuscript that Dr Carol Readhead and Dr Kate Hardy have given, as with much else, is deeply appreciated.

A PHOENIX PAPERBACK

First published in Great Britain in 1997 by
Phoenix, a division of the Orion Publishing Group Ltd
Orion House
5 Upper Saint Martin's Lane
London, WC2H 9EA

A CIP catalogue record for this book is available from the British Library.

ISBN 0 297 84116 5

Typeset by SetSystems Ltd, Saffron Walden
Set in 9/13.5 Stone Serif
Printed in Great Britain by
Clays Ltd St Ives plc

Contents

Laban's Sheep

When Jacob had completed more than sixteen years' work as Laban's farm manager, Laban finally asked him what payment he wanted. Genesis (chapter 30) recounts Jacob as saying that he wanted no payment but asked if, as a reward, he might have from Laban's huge flock of sheep those which were speckled black and white or ring-straked. These sheep were a small minority of Laban's flocks and not as well regarded as the apparently pure-bred whites. Jacob clearly knew that his father-in-law, Laban, was a fraudster and untrustworthy. His track record in his dealings with others was poor. Indeed, his name in Hebrew, 'Laban', means 'exceedingly white' because he was regarded as shining with wickedness. Nine years earlier Jacob himself had suffered from Laban's deceit on his wedding day. Then he was hoodwinked into a marriage with Laban's unloved and ugly elder daughter Leah, who was disguised under a thick veil. Jacob must have had a rude shock later that night as he expected to see the pretty younger daughter Rachel in his bed.

In fact, as soon as Laban agreed to let Jacob have the defective sheep, he had his other shepherds remove all the spotted and speckled ones to a remote and inaccessible place, a feeding ground three days' journey away from where Jacob was. Undaunted, Jacob then selectively bred speckled sheep from the remainder of Laban's herd, to which he had full access during his work. Ancient people believed that preconception and antenatal stimuli could influence the developing characteristics of a child. It is a familiar notion, and is still believed by many people around the world. While Laban's animals were drinking at the trough of water or copulating, Jacob showed them speckled whitish sticks at which he had whittled. The Bible reports that these sheep then gave birth to mottled offspring.

What is this story all about? Jacob was certainly no fool; moreover he was a highly skilled husbandry man, having worked with breeding sheep for fifteen or more years. He certainly must have known, from personal observation, that a visual influence at the moment of conception could have no bearing on the outcome of the pregnancy. The speckled sticks were simply a smokescreen to deceive Laban's sons and servants about what he was really doing.

Speckling is a recessive genetic characteristic. Carriers of recessive genes for a particular trait do not show the recessive features of that trait, unless they inherit that recessive gene from both their father and their mother. A single copy of the gene from one or other parent merely makes them a carrier of the trait, but they look externally in every way normal just as a non-carrier. In technical language they have a normal phenotype, but an abnormal genotype. Jacob had recognized, from his years of making Laban rich with his experience of sheep breeding, that the pure whites occasionally gave birth to sheep with abnormal speckling. By identifying the animals which gave rise to the speckled offspring, he could identify carriers (see Figure 1).

These he then selectively bred – the Bible tells us he segregated the animals which showed the most reproductive strength – and soon had a massive flock of speckled but entirely healthy sheep. Genesis records how he soon became richer than Laban with the flock he accumulated.

So humans have been manipulating reproductive processes since recorded time. This manipulation extended beyond animal husbandry to human reproduction. Ancient Egyptians used a variety of substances, including pessaries of crocodile dung, for contraception. Vedic literature is full of references to sex selection, a theme which carries on through ancient Greece, through Talmudic literature and more recently through medieval times. Although we think of eugenics as essentially a twentieth-century 'solution' for purifying the racial and mental health of society, the manipulation of populations started in the depths of history. The destruction of female babies on the hills of Sparta is relatively recent, but the ancient Hebrews attempts to exterminate all the Amalekites, and to prevent their reproduction, reflects a human compulsion since Biblical times. Indeed, the Bible is one of the richest sources of examples of manipulation of human conception. Onan 'spilt his seed' to avoid the obligation of marrying and having a baby by his childless sister-in-law – the earliest Biblical example of contraception. No fewer than three out of four of the matriarchs, Sarah, Rebecca and Rachel, were infertile, and various remedies from surrogacy to mandrakes were tried.

Modern ambivalence about our ability to influence the next generation therefore seems surprising. There is a popular notion that, at the turn of this millennium, mankind stands on a precipice. All our technology, it is frequently claimed, has overtaken us and now threatens us and our planet. Advances in nuclear physics, deforestation and expansion of the use of the internal combustion engine are all seen as frightening enough, but the new developments in reproductive technology are held to be the most dangerous. Presumably this is because they are so personal and question our very nature. Fertility treatments undermine conventional family structures; the selection of children for desirable characteristics such as gender menaces the economy of nations; cloning research imperils our species, yet contraception has failed to check the exponential growth of the world's population.

In this monograph I try to analyse where reproductive advances are taking us. Too frequently, attempts to make predictions are based on poor scientific understanding and inadequate assessment of the kind of technology likely to be available. I hope that the scientific indicators are both encouraging and reassuring. I do not share the alarm and pessimism which is often declared and I hope my analysis may persuade readers of the essential value and relative safety of the development of these advances.

Manipulating Fertility

Perhaps the greatest developments in human reproduction in the last two decades have been in the field of infertility treatments. Until very recently indeed, it was virtually impossible to treat most male infertility effectively, and after treatments for female infertility at least two-thirds of women remained childless. Moreover, probably half of those conceiving after treatment did so spontaneously. That is to say, the treatment itself probably played little part in the successful outcome.

The first major breakthrough came in the 1960s, when scientists managed to measure the hormones secreted by the ovary using techniques such as radioimmune assay. This meant that it was possible to see if a woman was not ovulating – the commonest cause of female infertility. Improvements in biochemistry also led to the isolation or synthesis of hormone preparations capable of stimulating ovulation, given either in tablet form (clomiphene) or by injection (menopausal gonadotrophins). Another very significant advance was the ability to pinpoint when ovulation was occurring. In this respect, ultrasound detection was revolutionary. Using ultrasound, doctors in the early 1980s could image the ovary and detect when the follicle containing the egg was about to rupture resulting in ovulation. This led to better timing of insemination or intercourse.

While improvements were rapidly happening in hormone therapy, laparoscopy was also being developed. Although it was possible to put a telescope into the abdomen and inspect the contents of the pelvic cavity before the Second World War, laparoscopy only really became established with the improvements in optics and the development of powerful halogen light sources which generated little heat. This revolutionary technique gave remarkably clear views of the uterus, tubes and ovaries and enabled surgeons reliably to detect the second commonest cause of female infertility, namely blocked fallopian tubes. Before the development of laparoscopy, very often the only way a fertility specialist could identify tubal disease was by performing a laparotomy, a major operation to open the abdomen and inspect the contents of the pelvis. Apart from the relatively huge trauma involved, and the long convalescence needed, this kind of exploratory procedure itself carried risks to a woman's fertility by sometimes causing adhesions to form between the tubes or ovaries and other tissues such as the bowel. Occasionally such surgical scarring could block or damage fallopian tubes which had previously been entirely healthy. So laparoscopy was an undoubted breakthrough. By 1976, some more adventurous surgeons were even starting to use the laparoscope to conduct simpler surgical procedures (keyhole surgery) to repair tubes which were blocked and divide adhesions. Such minimal invasive surgery became more popular with the development of medical lasers and fine electrosurgical instruments. Another significant advance was in microsurgery. The human fallopian tube is small and delicate – at its narrowest part, where it is frequently blocked by infection or other disease, it measures no more than 0.5 millimetres in diameter. The use of the operating microscope provided the only accurate means of removal of pathological tissue and rejoining blocked tubes reliably.

Treatments for male infertility – when low sperm count was the cause – lagged behind. By 1980, there were over 200 drugs which were in more or less frequent use to try to stimulate the testis to produce more sperm. The fact that so many different preparations were developed argues that none was very effective. Indeed, drug treatments for male infertility have now almost been completely abandoned because they were virtually useless except in a tiny proportion of men.

In Vitro Fertilization

The greatest breakthrough was the birth of Louise Brown in 1978 after *in vitro* fertilization (IVF). IVF was initially developed to bypass hopelessly damaged tubes when surgical reconstruction was a forlorn procedure. IVF using natural ovulation was limited because humans normally only produce one egg at a time. Thus drugs are given to stimulate the ovary to produce a large number of eggs simultaneously – perhaps ten or even twenty. These are then collected by sucking them from their follicles in the ovary when they are matured. As immature eggs are not capable of being normally fertilized, much research went into working out precisely when ovulation was about to occur, so that only eggs which were fertilizable were collected. These eggs are then exposed to sperm in dishes, and, once fertilization has occurred, the resulting embryos are grown in culture fluid, in an incubator. The incubator and the culture media allow an environment which loosely mimics conditions inside the body, so that normally some viable embryos are produced. After seventy-two hours in culture, two or three healthy-looking embryos can be selected for transfer back to the uterus.

Although IVF was originally used for patients with diseased tubes, it soon became clear that it could also be used in some cases of male infertility. This is because fertilization is facilitated in these artificial conditions when there are not enough sperm produced for natural fertilization. By the late 1980s, IVF was also being used for many other causes of infertility, including those patients for whom there was no clear-cut cause for the infertility.

Currently IVF is demanding. It requires quite massive stimulation to the ovaries with drugs, and this can cause considerable ovarian enlargement. The hyperstimulation which sometimes inadvertently happens can make a woman ill. Moreover, considerable monitoring for up to two weeks with daily hormone blood tests and regular ultrasound is needed to ensure that only mature eggs are collected. Such testing is emotionally tiring for the woman. Even with all this intervention, IVF has a relatively low success rate and in the best clinics only 20 per cent of treatment cycles results in pregnancy. Moreover, the need to transfer more than one embryo simultaneously to improve the success rate means that many IVF treatments have the unpredictable result of multiple pregnancy – twins, triplets or even quadruplets. This means that the risks of losing the entire pregnancy are greatly increased – humans were not designed biologically to produce a litter.

In the last five years, there has been a major advance in the treatment of male infertility. Even with IVF, many men producing poor sperm samples are unlikely to ejaculate sufficient normal sperm to provide a good chance of fertilization. There has been a genuine revolution with the advent of micromanipulation. Using sophisticated instruments and specially designed optics on microscopes, it has been possible to select an individual spermatozoon and inject it directly into the egg. This treatment, called Intracytoplasmic Sperm Injection, or ICSI, has been remarkably successful. Even those men producing very few sperm – perhaps less than 1 per cent of normal numbers – are now treatable. At the time of writing, some 4,000 babies have been born using this approach.

IVF still has a poor success rate – although treatments can be repeated – admittedly, with considerable expense and inconvenience. The latest statistics show that only about 14 per cent of treatments result in a live birth. There are several reasons for this. Firstly, it is clear that humans produce very few embryos which are capable of normal further development. This is why the average fertile couple tend to require several months of regular sexual activity before successful conception occurs. In Britain, it seems that about 18 per cent of couples will conceive in each menstrual cycle. Published reports suggest that American couples are slightly more fertile, with about 20 per cent conceiving each month. The most fertile are Australian women, with a cumulative conception rate of around 22 per cent per month. This may be because American and Australian citizens are more sexually active. In spite of their national reputation, the French do not do so well.

On average, only one in every five embryos produced by IVF results in a pregnancy after embryo transfer. All available evidence suggests that this is because many, if not most, human embryos have defects incompatible with viability. It is true that the uterine environment into which they are placed could be hostile, or incapable of providing proper support for implantation, but there is very little evidence for this. Detailed microscopic examination of human embryos shows that about 20–25 per cent have an abnormal number of chromosomes. In general, an abnormal complement of chromosomes means a major restructuring or deficiency of genetic material, incompatible with life. One exception is trisomy 21, three copies of chromosome 21, causing Down's syndrome. Although some Down's babies survive after birth, probably far more trisomy 21 embryos do not implant, or miscarry early. It seems that failure to allow survival *in utero* is a kind of natural safety-valve, guarding against large numbers of abnormal babies being born.

Chromosomal defects represent a fairly massive abnormality, affecting a whole number of different genes along the entire length of a chromosome. Many human embryos show other more subtle imperfections. For example, a majority of embryos contain some dead or dying cells. In some cases this cell death is probably programmed – the phenomenon is called apoptosis. Apoptosis occurs in all tissues and accounts for the constant remodelling that happens during development. This is how the web between the fetal fingers gradually disappears during growth; it is also how a tadpole loses its tail as it grows. But some embryos show excessive cell death during the very earliest stages of development, or just abnormalities of cell division. It is quite common for some embryos, for example, to have cells which contain more than one nucleus – or no nuclei at all. Many of these imperfections are not detectable with an ordinary light microscope when embryologists make their routine inspection of an embryo before its transfer to the uterus after IVF.

It is also likely that a number of human embryos do not develop because of as yet undetectable gene defects. For example, there are many genes which control early embryonic growth, implantation and development. If for some reason these genes are not properly expressed during early development, progression to the next stages of fetal life may be prevented.

The Future of IVF

Improved Hormone Treatments

The first development in the next few years will be improved methods of stimulating the ovary to obtain more fully mature eggs with greater efficiency. The hormones currently used are derived from the pituitary gland. This gland secretes follicle stimulating hormone (FSH) which stimulates the ovary *in vivo* to produce a mature follicle. Until recently this hormone was obtained by extracting it from the urine of menopausal women. After the menopause, very large amounts of this hormone are naturally made by the pituitary. It is an attempt by the pituitary to stimulate the failing ovary. During normal reproductive life relatively small amounts of this hormone are secreted. Once the menopause has commenced, the ovary is incapable of responding to stimuli, so the pituitary compensates by pouring out more hormone. The excess spills over into the urine, from where it can be concentrated chemically and extracted for therapeutic purposes. However, the body treats FSH extracted from urine as a foreign intrusion, and, when it is given to another woman by injection, antibodies can be formed. Also, biologically derived drugs produced like this often have varying strength, which means their effectiveness varies considerably from batch to batch.

One of the advances in modern therapeutics has been to synthesize these, and similar protein hormones by genetic engineering. It is now possible to produce hormones normally made in the pituitary gland entirely artificially. Cells taken from the ovary of Chinese hamsters are grown in culture vats, and these cells multiply vigorously in a kind of soup. Before commencing active growth, the cells have been modified with the human gene which makes FSH. Once the 'soup' is really brewing vigorously, massive amounts of hormone are produced relatively economically and very reliably. Moreover the process of production can be tailored, so that the active principle can be altered by changing its molecular structure. The 'designer drug' recombinant (see Glossary) FSH is likely to be more potent than biologically produced hormones and over the next five years there will be a new class of more effective drugs. It will be possible to stimulate the ovary more precisely and with few side-effects, and less chance of hyperstimulation with its unpleasant and dangerous side-effects.

In Vitro Maturation of Eggs

A much more important advance is also under way. It may be possible to dispense with the use of expensive stimulatory drugs altogether, using immature eggs recovered from the ovary. There are, apart from any other consideration, good commercial reasons why this would be helpful. IVF is an expensive treatment, and generally costs £1,500 to £2,000 per treatment cycle. This cost includes monitoring the time of ovulation, the removal of the eggs and the laboratory expenses. It does not include the cost of FSH, which is itself expensive. In young women, the added cost of the drugs for a single course of treatment is typically £200–£500. Women in their thirties or early forties usually require a much more hefty stimulus and the extra cost of drugs can be more than £1,000. These factors make IVF a privileged treatment, affordable only by the relatively well-off, or those fortunate few who are able to obtain public funding (such as National Health Service support) for their treatment.

Using immature eggs recovered from the ovary could revolutionize the whole of reproductive medicine within the next ten years. It is feasible because a woman is born with her entire complement of eggs already formed in her ovaries. Very early in intrauterine life, a female fetus has about five million eggs formed in her ovaries. Each of these eggs is genetically unique, containing the individual mix of genes from mother and father. Remarkably, most of these eggs start to atrophy even before birth so that, by the time a baby girl is born, she is left with about two million eggs. By puberty, there are possibly between 200,000 and 300,000 left. Thereafter, only one egg each month will actually be ovulated. Over a thirty-year period therefore, some 400 will reach normal maturity and, given the average-sized family, two or three will develop into babies. This extraordinary redundancy seems all the more remarkable when one considers that, as each egg is genetically unique, the inherited characteristics of our children are left purely to chance.

All these eggs are contained in tiny primordial follicles in the surface of the ovary, called the cortex. This is a skin, just a few millimetres thick, surrounding the ovary. In any average young woman, one square millimetre of this cortical tissue will contain between 200 and 400 eggs. Being on the surface of the ovary, it is extremely accessible to a surgeon. A simple needle stab under local anaesthetic will collect a small sliver of tissue containing several hundred eggs. This tissue can then be frozen in liquid nitrogen and stored. Methods are being developed for the successful thawing of this tissue and for separating from it individual follicles containing eggs. Work now in progress will eventually allow such follicles to be matured in artificially produced culture media (maturation *in vitro*) to the stage where the egg can be fertilized. The hope is that embryos formed in this way will be normal and have a chance of survival.

The main problem facing researchers studying maturation of follicles *in vitro* is that we are uncertain which genes and which growth factors are essential. It has already been possible to mature follicles to the early antral stage, the stage at which the egg prepares for subsequent fertilization. This has been achieved by dosing the culture media with minute quantities of the hormone FSH, together with other factors which are guessed to be important. It is likely to be only a matter of time before scientists will be able to collect fully mature eggs in such systems.

This would be a major breakthrough in fertility medicine. Eggs could be produced cheaply and there could be a plentiful supply. It could provide a source of donor eggs for those women with no eggs of their own, having suffered a premature menopause. It would also be highly useful for cancer victims. Many cancers are quite common in relatively young women, such as leukaemia, lymphatic cancers and some forms of breast cancer. One problem that these young women face is that the chemotherapy and the radiotherapy needed to cure their cancers make them sterile by destroying egg cells in the ovary. In future, before undergoing treatment for cancer, such patients could have small amounts of their ovarian tissue stored. Once the cancer has been cured, they could have IVF with their stored, matured eggs and a genetically related child. But the biggest opportunity will undoubtedly be the possibility of replacing the cumbersome treatment that women now experience during routine IVF.

The continuous process of egg loss from the ovary during reproductive life means that women become increasingly infertile by a process of attrition. A woman's fertility halves during her early thirties, and women over forty are likely to be severely infertile. In most of Western society, women are delaying pregnancy for many good social reasons. Sadly, many women find that by the time they are in a position to conceive they have left things too late. A substantial number of women attending infertility clinics are infertile simply because of this natural ageing process. Nothing at present can be done for them beyond offering them donated eggs from another individual. The future could bring extraordinary procreative liberty. Is it too far-fetched to envisage the time when more mature women use eggs kept in the deep freeze for twenty years? I like to think of the example of the young law student, who may take her final examinations and obtain her university qualification at twenty-one years old. Having had some ovarian tissue stored, she could read for the bar, find chambers and develop a career as a barrister. Once she has taken silk in her early forties, she could return to have her eggs matured in the laboratory. After fertilization with her husband's spermatozoa, the embryos could be transferred with a simple ten-minute procedure. Such a process would require less monitoring and drug use, and would be cheaper and much less emotionally demanding than are current IVF treatment methods.

Genetic Diagnosis

One of the most challenging developments in IVF has been methods to screen human embryos for genetic defects. There are about 5,000 single gene defects. They frequently cause death in young children and those gene defects which are not fatal generally cause severe disability.

Gene defects can be divided into three groups. The most common are the recessive genetic defects. These cause a problem only if both parents carry the same defective gene and both pass that gene to the child. In such families, with both parents carriers, there is a 25 per cent chance of any child being affected. Typical common recessive defects include cystic fibrosis, which causes severe lung and digestive disease, and thalassaemia, a blood disorder causing severe anaemia. About 500,000 babies worldwide die from thalassaemia each year. It is common in Britain in many immigrant populations, particularly some Asian families and those from the Mediterranean – for example, Cyprus.

The second group of gene disorders are those caused by dominant genes. Any person inheriting one copy of such a gene from either parent will suffer from the disease. It therefore follows that the parent carrying that gene will also have the disorder. Consequently, most dominant defects die out because gene defects tend to affect people from childhood onwards, and carriers rarely live to mature life. The dominant gene defects which are prevalent are mostly those which only express in older adults. The commonest is Huntingdon's chorea, a crippling neurological disorder which only starts to take effect normally when a person is over forty. By that time they may have had several children without realizing they have inherited this defect.

The last group of gene defects are those which are sex-linked. These are defects produced by genes on the X chromosome. Every female has two X chromosomes, males have only one. A woman may carry one of these defects on one of her X chromosomes, but because she has another normal X chromosome, she will not suffer the disease. A man is less fortunate if he inherits the 'wrong' X chromosome from his mother. Having only one X chromosome will mean that this defective gene, if present, will be expressed, and he will suffer from the disease. The chances of any man inheriting an X-linked disorder from a carrier mother are 50 per cent. There are about 300 known X-linked disorders. Among the commonest is Duchenne muscular dystrophy, which causes slowly developing paralysis in young boys – most dying from suffocation by their teens. Another is haemophilia, the famous disorder affecting members of the Russian Royal Family, which causes severe haemorrhage or bruising after slight injury because of the failure of the blood to clot.

Families generally only know they carry a gene defect if there is a family history. Most frequently, the majority of couples find they may be carriers when they have a child who turns out to have one of these horrible diseases. In the long-distant future more and more people may undergo routine genetic screening for the commoner disorders and therefore will know whether they are at risk. It seems extremely unlikely that there will ever be adequate screening for the rare disorders and most individual gene disorders are not very common. It is just that there are so many different ones that collectively they present a significant medical problem. Until recently, the only practical way of dealing with these defects has been to screen pregnancies, once the pregnancy is well established. This requires amniocentesis or removal of a piece of tissue from the placenta, and both procedures are not entirely without hazard to the pregnancy. Once a diagnosis has been made, the couple may face the heartrending decision of whether to have an abortion.

A major development occurred in 1990 with the technique of preimplantation diagnosis at the Hammersmith Hospital in London. Using IVF, embryos were obtained from couples who had already had a child die of a genetic disorder. Three days after fertilization, when the embryos comprised just eight cells, microsurgery was used to remove a single embryonic cell. This cell was frozen and immediately thawed to release the DNA contained in the cell's nucleus. The DNA was then examined to see if the embryo from which the cell came carried the specific defect in that family. Initially, when the technique was first started, it was decided to concentrate on diagnosing just the sex of the embryo when the woman was a carrier of a sex-linked disease. Determination of DNA associated with genes for sex was thought at the time to be easier than looking for a specific gene disorder. The strategy was to transfer female embryos only to those mothers who had lost a male child with muscular dystrophy, adreno-leukodystrophy or severe X-linked mental with physical retardation. Three women were initially treated and each had two female embryos transferred. Two became pregnant and both had twin girls. By 1992, the technology was extended to make specific diagnosis of an individual gene defect. Cystic fibrosis was chosen as being the most important because it is so common – about one in 2,000 babies are born with this disorder. Since that time about seven or eight different genetic disorders have been detected in embryos by these methods. A number of healthy babies have been born to parents who had previously watched a child tragically die. These families chose preimplantation diagnosis after IVF, with all the rigours involved, because they were reluctant to consider termination of an established pregnancy. Their decision was often based on their moral views about the abortion of an established pregnancy.

Preimplantation diagnosis has now expanded to include chromosomal disorders. The breakthrough became possible when scientists learnt how to stain the DNA in individual chromosomes, or parts of chromosomes, with brightly coloured fluorescent dyes. Gene defects are not the only problems which cause genetic disease. Less than half of 1 per cent of all babies born have a single gene defect, but a defect of one or more chromosomes is much more common. Indeed, at least 20–30 per cent of human embryos have abnormal chromosomes, one of the commonest causes of failure to conceive or of miscarriage. Down's syndrome, three copies of chromosome 21, is one of the few disorders compatible with viable life. Nonetheless, most of these children are severely handicapped usually with profound mental retardation and other defects such as an abnormal heart.

Chromosome staining (or painting as it is frequently called) is now being used to detect these severe abnormalities in embryos, from those women who have had repeated embryonic loss or early pregnancy failure; or who have merely repeatedly failed IVF. One prospect for the relatively near future is that, because many older women have eggs which are likely to be more prone to chromosomal abnormality, this technique can be used for older women having IVF treatment. It may just be possible to improve the success rate of IVF in these women quite dramatically. Biopsy and screening of cells taken from the embryo during the first five days after fertilization may ensure that doctors select healthy embryos for transfer.

However, at present, chromosomal detection of this kind is limited. This is because it requires the complexities of operating on the embryo to remove cells first. Microsurgery of this sort can damage embryos and prevent subsequent development. Moreover, the technique is time-consuming, labour intensive and consequently expensive. Because it is difficult to see how it could be automated, it is unlikely to be suitable for the routine screening of large numbers of human embryos.

In the next few years, preimplantation genetic diagnosis will almost certainly be applied to screen those embryos from women who are at risk of having disorders which only express themselves in later life. Typical of these are the cancer genes. Familial breast and ovarian cancer, polyposis coli (which causes bowel cancer) and some forms of brain cancer such as retinoblastoma are diseases which are inherited. A carrier of one of these genes is highly likely to develop the cancer at a young age. Moreover, many of these cancers are hard to treat because they grow so rapidly and spread to other organs. For example, even if a breast cancer from one of these victims is successfully treated by radiotherapy or surgery of one breast, there is a likelihood of another similar cancer developing in the other breast at a later date because of the overall genetic predisposition.

This approach to preventing cancer raises increasingly interesting ethical problems. Perhaps the main one is that people who carry these genes will be perfectly capable of conducting full and useful lives up until the time when the cancer strikes. Should we therefore be destroying embryos who may survive until the age of forty or even fifty years old? Franz Schubert was thirty-one when he died of syphilis, Wolfgang Amadeus Mozart died at thirty-six, and the poet Keats contracted tuberculosis as a teenager, dying from lung haemorrhage at twenty-five. All contributed immeasurably to human culture and happiness. Who is to know that we might not be selecting out another Schubert by discarding a particular embryo? This argument seems a poor one. It could equally be true that the embryo free of the gene defect could be the one with the talent of Schubert, or indeed a Picasso, who lived until his eighties. The characteristics which contribute the nature of an individual are always likely to be undetectable by these methods and therefore such selection would always be random.

Currently preimplantation diagnosis can be done by examining individual genes, or by staining all or part of some of the chromosomes. As already mentioned, both methods are complex. There is potentially a third option. It is now clearly established that many of the genes which contribute to the general function and health of an adult are already 'switched on' and working during very early development. These genes are often producing proteins which may be detectable without necessarily making a biopsy of the embryo.

A good example of this is the enzyme HPRT, or Hypoxanthine Phosphoribosyl Transferase. Absence of this vital enzyme is caused by mutations in the gene responsible for its manufacture. Children who are born with deficiency of the HPRT enzyme suffer from the terrible disease Lesch-

Nyhan syndrome (see Glossary). The enzyme deficiency causes, among other things, mild mental retardation and very severe cerebral palsy. Such children suffer from constant writhing movements and are spastic – for example, they cannot walk. Most horrific is their compulsive desire to self-mutilate and consequently they have to be restrained, strapped in a wheelchair. Without such restraint, these children sometimes bite off the tips of their fingers, for example. Even with restraint, many of these children mutilate themselves by biting off their lips or tongue. In spite of the most thorough care, it is rare for these affected children to live beyond the age of about twelve.

It may just be possible to treat these children by gene therapy. There have been a number of attempts to introduce the unmutated gene into the child's white blood cells, so that sufficient HPRT is made to alleviate this horrible disease. A discussion of somatic cell (see Glossary) gene therapy is beyond the scope of this book, but suffice it to say that, given our current state of knowledge, such gene therapy is really only possible once the disease is firmly established, by which time the severe neurological damage is irreversible. In the absence of effective treatment, techniques such as preimplantation diagnosis seem worth exploring.

In a very few cases, preimplantation diagnosis has been used to prevent Lesch-Nyhan by screening embryos for the specific gene mutation and then transferring only healthy embryos to the uterus. But this, of course, has involved a biopsy. To avoid this invasion, attempts to try to detect the presence or absence of the gene product HPRT have been made. The hope was to examine the fluids in which the embryo is grown in culture to see whether this enzyme has been secreted by embryos from carrier mothers. Such a technique has the beauty of being entirely non-invasive – that is, it involves no surgery. Theoretically it is also likely in time that measurements of such enzymes could be done using automated machinery. This would eliminate much of the intensive work that is currently needed both for embryo biopsy and for detection of specific segments of DNA, or of parts of chromosomes.

Unfortunately, this non-invasive approach – a general strategy which could be applied to a good number of different hereditary diseases – is not yet possible. The major problem is that most of the gene products of interest are already present in the substance of the egg before fertilization. This is because the mother, being unaffected, has normal genes. Some of the products of maternal genes linger in the embryonic cells which are formed from the cytoplasm of the egg, and they may last for several days. However, preimplantation diagnosis using chemical screening in this way may be feasible eventually. One strong possibility is that embryos will be grown for longer in culture until the blastocyst stage, just before implantation. The human embryo becomes a blastocyst on about the fifth day after fertilization and normally comprises well over fifty cells. Examination of the blastocyst and the secretions in fluids around it would be advantageous; because there are more cells by this stage, there is more metabolic activity and consequently there should be more of the specific gene product available for testing. This would mean that the diagnosis could be made with more security. Secondly, by this stage of development, it is much less likely that any genetic 'residue' from the mother would still be present to confuse the diagnosis.

The current drawback to testing embryos in culture is that it has been surprisingly difficult to grow human embryos outside the womb for more than the first two or three days of life. Blastocyst culture is currently problematic because, as yet, we do not fully understand the ideal milieu needed for the embryo once it starts to grow to later stages. This is a problem which, as we shall see later, is almost certainly soluble and which when solved will produce substantial other benefits.

Selection of the Best Embryos for Transfer

In vitro fertilization has a very low success rate unless more than one embryo is transferred to the uterus simultaneously. This is because, as has been previously described, most human embryos seem incapable of advanced development, possibly due to a variety of defects. It is this essentially biological problem which has led doctors to transfer several embryos to the uterus simultaneously. While this practice increases the chance of a pregnancy with at least one embryo implanting, it does risk multiple implantation – which carries risks of miscarriage and multiple births. Since the advent of IVF, the incidence of multiple birth has more than doubled in Britain. Most triplets, and nearly all quadruplets, that are born have been conceived following assisted reproductive treatments of one sort or another. Although high-order multiple birth has often been seen as a triumph in the lay press and by some of the public, it is little short of disastrous in the majority of cases. Apart from the difficulty that parents have in coping with many babies after birth, there are significant medical problems. Firstly, triplets virtually never go to term; indeed, most are born six to eight weeks prematurely, if they survive pregnancy at all. Having three or four babies develop simultaneously in the womb greatly increases the hazards of pregnancy, with high blood pressure, toxaemia, diabetes and blood clots being much more likely. Delivery by Caesarean section is nearly always needed and most babies born this premature require intensive nursing in incubators, often for several weeks. It costs the National Health Service approximately £500 to £1,000 each day to care for a very small baby intensively, and therefore the cost of supporting premature quadruplets can be very substantial. Moreover, surviving premature babies are far more likely to have developmental and health problems later. For all these reasons it is hardly surprising that *in vitro* fertilization has not always been a very popular treatment with paediatricians and health service managers.

Unfortunately the current imperfect technology puts much pressure on infertility specialists to continue transferring several embryos simultaneously, in order to increase chances of pregnancy. IVF is a costly process and most patients simply cannot afford to go through endless cycles of treatment. In Britain the regulatory authority (HFEA) has placed a limit of three embryos at once, but even this carries quite a risk of multiple birth. In most other countries there is no statutory limit and it is not uncommon for practitioners to transfer five, six or even seven embryos.

A major advance will clearly be tests on embryos in culture to ascertain which are most likely to be viable and to implant. The eventual aim is to give IVF a high chance of success with just single-embryo transfers. So far a number of tests have been devised, but none is very reliable.

One of the best of these tests has been non-invasive assessment of metabolism. Remarkably, it is actually possible to measure how much energy an embryo is using. The idea behind such measurement is the assumption that the more energy used, the more likely the embryo is to be growing actively and therefore to be viable. The technique is relatively simple. Each fertilized egg, invisible to the naked eye, is placed in a tiny droplet of culture medium so small that if spilt on one's finger one would hardly feel wetness. To remove the droplet from outside influences, it is placed under a globule of sterile inert oil. The medium in which the embryo is allowed to grow contains a measured amount of sugar. After twenty-four hours, the embryo can be removed from its droplet and placed in a new, freshly prepared one. The original droplet can then be analysed to see how much sugar has been taken up, and thus the amount consumed by the embryo in twenty-four hours can be estimated.

Measurements on several hundred embryos have consistently shown that embryos more likely to be viable consume more sugar and are more metabolically active. Of particular interest is the finding that male embryos are, on average, 18 per cent more active than female embryos. It seems that we men are more aggressive from the moment of conception. Unfortunately, sugar consumption is not reliable as a discriminatory test. There is so much variation in metabolic activity of this sort from embryo to embryo, irrespective of sex, that the test is insufficient to determine which embryos are most likely to become a baby if transferred. One problem is that sugar metabolism is a very basic process, and even cells which are about to die may consume considerable amounts of sugar.

In the near future, however, there are likely to be significant advances. Although sugar metabolism is a very basic function, there are other forms of metabolic measurement which may reveal more. A promising area is to see what amino acids are taken up into the embryo during growth, using similar culture techniques. Amino acids are the vital components of proteins. Proteins are the essential building blocks during growth and are needed for cell division. Many amino acids are needed for embryo culture and it should be possible to measure their depletion from culture media over a fixed period of time, thus gauging an individual embryo's growth potential. Continued research in this area should make it possible to select viable embryos with rather more accuracy, and it is likely that this approach may be applicable within the next five years.

Metabolic measurement is not the only refined, non-invasive assessment likely to give information about embryonic viability. Cells that are actively dividing need substances called growth factors. Growth factors are chemical messengers which attach themselves to receptors on the cell wall. Attachment to receptors leads to the generation of further messages within the cell which tell it to grow and divide. It is already known that embryos require growth factors of various sorts to stimulate their growth and researchers are assessing which growth factors are most important, and which may be needed to provide the best culture environment. It is also true that the embryo itself may produce different growth factors as its own genes are switched on. These growth factors are messengers which tell the lining cells of the uterine cavity that the embryo is present. It is likely that these messages are important in preparing the endometrium (see Glossary) for implantation. Consequently the most viable embryos, the ones most likely to implant, are most likely to produce good amounts of growth factors. In due course, their measurement is likely to be quite simple even though they are present only in low concentrations. It is more than probable that, within the next ten years, these measurements may help pick the embryos most likely to be healthily viable.

Improvements in Culture Media

It is clearly established that embryos of different species require different conditions for the best growth. For example, most embryo culture media have traditionally contained glucose. This is, on the face of it, hardly surprising as glucose has always been regarded as a basic energy provider for most mammalian cells. It now turns out that glucose, even in moderate amounts, actually tends to prevent growth in mice embryos in the first two days after fertilization. The evidence that glucose is 'poisonous' to early human embryos is not nearly so convincing. However, it is clear that the human embryo culture media in which embryos are currently grown after IVF are less than ideal. The evidence for this is that human IVF embryos grow more slowly than after natural fertilization and normal incubation in the fallopian tube. Secondly, they tend not to survive very well if grown more than three days in culture media outside the body. Thirdly, embryos which do survive usually have fewer cells than embryos at an equivalent stage after natural fertilization and growth inside the body. It is also true that, in most IVF units, any single embryo transferred to the uterus after IVF on or around the second day after fertilization only has around a 10 per cent chance of becoming a baby. In units which take a great deal of trouble over their laboratory techniques, the chances of a single embryo going on to further development is better than 10 per cent, reaching a maximum of around 20 per cent. All these observations lead to the conclusion that the artifical conditions in which embryos are kept in culture laboratories are not optimal.

The gaps in our knowledge about the best environment for human embryo growth will undoubtedly be filled within the next five to ten years. This is one of the most reassuring developments in IVF because, once we know the best solutions in which to nurture human embryos, there is likely to be a substantial improvement in treatment. Currently, embryos kept in culture in most laboratories seem to deteriorate. There would be a great advantage in being able to culture the embryo to the fifth day, when it becomes a blastocyst, because available knowledge suggests that any single normal blastocyst transferred to the uterus would have at least an evens chance of developing into a healthy fetus. Given that a three-day embryo only has a 10 to 20 per cent chance of viability, the availability of good blastocysts could give massive improvement to current results.

There are several promising ideas already being explored. It is now clear that the right concentration of carbohydrate can promote human embryonic development and the wrong concentration can inhibit it. We do not yet know what is ideal for the human embryo, but its requirements are likely to change during development. It seems that pyruvate is the essential carbohydrate needed for the first two days, but thereafter the human embryo needs glucose. It may well be that the culture media in which embryos are grown need to be changed depending on the precise stage of development.

While there has been quite a lot of research on the

energy requirements of embryos, much less is known about their need for amino acids, the building blocks making essential proteins. There are twenty common different amino acids and their relative importance to developing embryos is a mystery. One of the most interesting, and possibly important, is the substance glutamine. Some embryos require glutamine for best growth. This requirement varies not merely between different animal species, but even between different strains of the same species. Most mouse embryos, for example, need it, but there are some strains of laboratory mice whose embryos can develop normally in media which contain very little or no glutamine at all. New research in the human suggests that the presence of glutamine in the culture media may greatly enhance growth in the early stages. Currently, as conventionally used media in human IVF laboratories do not have this substance, there is a very strong case for adding it – but in the right proportions. Too much, of course, may inhibit growth. I use glutamine as an example, but there is a whole range of other compounds, including inorganic metals and various salts, which when added to media may possibly make a considerable difference to the clinical outcome of IVF.

One very new and exciting area is that of growth factors. We now have evidence that embryos almost certainly need these compounds for best development. They are present in varying quantities in the fallopian tubes and uterus, and seem to have a most important influence. So far no laboratory has tried adding these substances systematically to embryos in culture. Some preliminary work in our own unit suggests that the addition of one particular factor, from the family of insulin-like growth factors, may actually enhance development by as much as 30 per cent. Undoubtedly, therefore, further research in this area will yield rich rewards for infertile patients.

Sex Selection

Since time immemorial, humans have tried to influence the sex of their babies. It seems males were preferred to females in most ancient societies. An exception to this is possibly the dilemma of the Hebrew slaves in captivity in ancient Egypt. The Bible reports that Pharaoh ruled that all male children born to them should be drowned, so that only females would survive, thus reducing the chance that the Hebrews would rise in revolution. This is why Moses' mother hid him and floated him down-river in a reed basket.

The ancients tended to believe that boys were generated from the right side, girls from the left. Coital position was therefore considered important to influence the humours forming the fetus. Timing of intercourse during the menstrual cycle, delaying orgasm, changing the amount of salt in the diet or the acidity of the vagina, taking up less stressful living, have all at times been favoured methods of influencing the sex of the fetus. There is little evidence for the efficacy of any of these remedies. While the remedies themselves are not particularly notable, the fact that they have been employed in ancient Egypt, ancient Greece, Roman times and many other civilizations in almost all parts of the world ever since records began argues that most peoples found little moral objection to the notion of sex selection.

The sex of a baby is determined by the sperm fertilizing the egg. Sperm carrying the Y chromosome produce males. The Y chromosome is physically smaller than the X chromosome, and there is therefore a minuscule difference in the weight of male- or female-bearing sperm. Some authors have argued that male sperm swim faster and are also smaller. In spite of various scientific papers still published on the subject, there is no convincing evidence for the agility or increased speed of Y-bearing sperm. Nevertheless, this has not prevented commercially minded doctors and scientists attempting to devise methods to separate X- and Y-bearing sperm. These methods have the object of producing enriched sperm samples for artificial insemination into the female partner.

Sperm separation generally involves passing sperm through a fluid of relatively high density, and/or spinning samples in a centrifuge to isolate those of a particular weight. One popular method devised by a Dr Ronald Ericsson, and used at various times by one rather notorious clinic in London, employs these basic techniques, and the service is sold to quite large numbers of people who may give a preference for having boys. Without going into detail, there is little or no genuinely independent evidence that this – or indeed any other related method – actually works. There is therefore the possibility that a number of patients who come to such clinics could be hoodwinked or exploited.

At the present time, the only method of sex selection which definitely has a high chance of working is the one devised at Hammersmith Hospital, used for preimplantation diagnosis (see p. 16). This is very reliable, but it is complex – involving, as it does, IVF. Moreover, it is expensive and requires the biopsy of, and possibly damage to, an embryo. For this reason, the UK regulatory authority has, probably justifiably, sanctioned its use only for those patients at risk of having a baby with a severe sex-linked disorder.

There is no doubt that in the near future, perhaps by the year 2000 or 2002, there will be a very reliable way of sex selection which will not involve the complexities of IVF, let alone embryo biopsy. Work in Cambridge in cattle clearly indicates that it is possible to enrich sperm samples with X- or Y-bearing sperm by using the method of flow cytometry. This involves placing a fluorescent dye on the X or Y chromosome, following which the sperm are sorted in a laser beam. There is doubt whether the fluorescent 'tagging' of sperm might carry genetic hazards, with damage to the DNA. It is possible that it does not and it seems that, with refinements, this kind of approach is likely to be used eventually in humans. While the apparatus and the technical back-up needed to do this are expensive, it is bound in time to be very much less involved than IVF. This is because, once the sample is enriched, only artificial insemination is needed, the simplest of reproductive treatments.

This technology will open up a huge ethical debate. While it is feasible to regulate sex selection using IVF methods – existing legislation, together with the difficulty of treatment, ensures that – a simplified method involving sperm separation and insemination will be almost impossible to police. This will be particularly true in developing countries which have relatively sophisticated technology and where couples are under considerable social pressure to have a child of a specific gender – usually a boy. The lid may be kept on the technique for a while in Europe, but people in countries like India and China may well provide the impetus for the regular use of these treatments.

There are reasonably clear arguments against the widespread implementation of sex selection. Firstly, there is the concern that there would be a risk of changing the balance of the population, probably by increasing the ratio of boys to girls. Although this in itself seems highly undesirable, there is little clear evidence that it would bring about deleterious social pressures. A society filled with males may be more aggressive, but this is doubtful. Moreover, there is a strong possibility that, if males predominated, females may become more 'valuable' – and there would be pressure, both natural and social, to restore balance. Another argument against sex selection is that it would be likely to increase inequality in society, by 'benefiting' the better-off families. Should there be a preference for males, females may find a reduction in their status. Sadly, given that this is probably true of many contemporary societies, there is little evidence to suggest that sex selection would make matters any worse. One prevalent argument is that this kind of social engineering is ripe for state exploitation. Governments may use it to further interests not necessarily to the advantage of all members of that society, or countries beyond – for instance selecting in favour of males in order to create armies, or females in order to increase population.

Introducing New Genes

The most fundamental advance in reproduction this century has almost certainly been the development of transgenic technology. Using the access to embryos that IVF provides means that new genes can be introduced into them, thus changing their genetic expression as they grow older. 'Foreign' genes (or transgenes) were first injected into mouse eggs in the early 1980s. Among the early mouse experiments were those by Dr Marston Wagner and colleagues, who introduced the human gene which makes part of the haemoglobin molecules in red blood cells. At about the same time Drs Richard Palmiter and Ralph L. Brinster were introducing growth human genes into mice embryos – making the so-called giant mouse. The mention of a giant mouse may conjure up surreal images, but it was in reality just a slightly larger-than-average mouse, which grew faster and made more efficient use of foodstuffs. Mice remain the most common species for scientists studying the effects of transgenes, and such animals have become very important models for human diseases with a genetic component, and for new therapies. They have been very valuable for studying cancer-causing genes and are used for the study of DNA and its function.

Making transgenic animals is not easy at all. It is now most frequently done by injecting a solution containing a number of copies of the gene of interest into the fertilized egg, shortly after the sperm has entered it, and before it starts to divide for the first time into two separate cells. While there are other stages of development when transgenic animals can be made, this is the stage when researchers have had the most success. Once the gene has been injected, the fertilized egg is returned to the tube or uterus of a foster female animal previously prepared for pregnancy by giving hormones. The process is technically demanding – for example, the glass needle which is used to inject the gene solution into the microscopic egg is so fine that its tip cannot be seen with the naked eye. The slightest untoward pressure on it breaks it, rendering it useless.

Normally genes injected into embryos this early in life will tend to enter all tissues, including the newly formed eggs or sperm of the fetus. Consequently, transgenic animals, when they become mature, will give rise to offspring which will also be transgenic. This potentially offers scientists a number of animals suitable for scientific study.

Injecting new genes is difficult enough, but, at present, even after successful injection the results are extremely unpredictable. Very often the fertilized egg is destroyed by the injection process. Hardly less frequently, the embryo develops lethal damage and does not develop. Most injected embryos that do develop and are subsequently born show no evidence of the injected DNA and are therefore not transgenic. Those that are transgenic may not express the gene of interest – that is, the DNA may be present in the animal's chromosomes, but it is not working. Some animals show expression, but only for a short period and, most frequently, the expression varies so that the DNA does not produce its full normal effect. Many transgenic animals are mosaic – that is, the gene is present and expressing in some cells or tissues, for example the liver, but not others, such as the brain. This can be very frustrating if the tissue of interest to the scientific team is nervous tissue, but not the liver. Injecting new genes in this way can sometimes unpredictably cause problems with other genes, perhaps by displacing them or altering their function. Moreover, transgenic animals may show serious congenital abnormalities. One expert scientist, Dr Carol Readhead, who has spent years making transgenic mice, tells how in one particular transgenic experiment she needed to inject and transfer around 1,600 mouse embryos to get a single young animal clearly expressing the gene that she needed to study.

There may be a number of ways of improving the poor success rates involved in making transgenic animals. One exciting idea which is being tested, and which would revolutionize transgenic technology, involves targeting sperm cells with new genes rather than embryos. The testis is making new sperm cells all the time during reproductive life. These sperm cells are derived from primitive stem cells, called spermatogonia. It is possible to remove spermatogonia from the testis and place them in culture. Once in culture, there are a number of ways in which new genes can be encouraged to enter these cells. One method involves attaching the transgene of interest to a suitable virus. Once in culture, the viruses infect the spermatogonia cells and release their own DNA and the DNA which has been piggy-backed with them. These new genes may then become incorporated into the spermatogonia. In any culture, only a certain number of cells will become invaded (or transfected) by the new gene, but there are various ways of assessing which cells are transgenic and sorting them away from the unsuccessfully transfected cells. These modified spermatogonia can then be returned to the testis. To prepare the testis for receipt of these new cells, it will have first been depopulated of all non-transgenic sperm, possibly by X-radiation or by the use of chemotherapeutic drugs which destroy growing sperm cells. Consequently, after transfer of the new spermatogonia, that animal will only produce sperm which have been modified. The beauty of this technique would be that transgenic animals would be much easier and far cheaper to produce, simply by natural mating – rather than by complicated embryo manipulation. This technique would also have the advantage that it could be applied to large animals. Currently, transgenic technology using embryos is so cumbersome that really only mice are generally used. For example, it costs over £25,000 to make a transgenic pig, suitable for one human cardiac transplant. This technology could be applied cheaply to a whole range of large animals such as cattle, sheep, pigs and monkeys.

However, manipulation of embryos is still likely to be the main way of producing transgenic animals for a considerable time. There will be progress in our ability to insert new genes into any animal embryo. Advances in molecular biology will also increasingly ensure that any gene inserted will work normally and express, and that this expression will be permanent. Great strides are already being made in the field of gene insertion for gene disorders and for the treatment of cancers. In humans, so-called gene therapy has been limited to inserting genes into somatic cells – cells in tissues such as liver, muscle, nerve cells and the cells which circulate in the bloodstream. Somatic cell therapy only has implications for the patient treated, and not for any offspring, because these inserted genes do not enter sperm or egg. Germ-line therapy – insertion of genes into embryos or germ cells (that is, eggs or sperm) so that these genes are inherited – is likely to be feasible in future. If it is possible to make transgenic mice, it is surely possible to make transgenic humans.

There are essentially two reasons for wishing to introduce new genes into human embryos. The first would be to correct gene defects which occur in that family, so that future generations would not suffer the particular disease carried in the family. Ultimately there is the notion that specific gene defects could be permanently eradicated. A second objective, perhaps further into the future, would be to introduce genes giving specific characteristics which are regarded as desirable. This concept leads to the idea of the 'designer baby'.

Not surprisingly, the idea of genetic engineering has provoked many expressions of horror and alarm. Indeed, the fear of such genetic manipulation has been so great that this was probably one of the most frequently used arguments in recent years against the development and use of IVF as a fertility therapy. For example, in the 1980s many Parliamentarians, theologians and journalists saw this as the main reason to ban embryo research completely. It is therefore worth analysing carefully the advantages, disadvantages and dangers of human genetic manipulation in embryos.

The potential advantages of germ-line therapy for disease treatment are not inconsiderable. Even though the technique is difficult, insertion of genes into embryos is likely to be easier than insertion into somatic cells. To date, numerous attempts have been made to insert 'corrected' genes into the tissues of sufferers from inherited diseases such as cystic fibrosis, but results are poor and few workers have been able to get these inserted genes to work for more than a few weeks. Moreover, because somatic gene therapy is potentially dangerous because it may induce cancers and cause other unexpected effects, it is really only considered once the disease is advanced; there are much safer and gentler treatments in the early stages of such genetic diseases. Correction of the cystic fibrosis gene in an embryo would have the huge theoretical benefit of treating the disease before it appears – consequently there would be no suffering. Secondly, the effects are more likely to be permanent, as experience with transgenic animals has shown. It is possible to get complete integration and normal function of inserted genes in some cases. Thirdly, of course, gene insertion into the embryo does not only prevent that individual from getting the inherited disease, but prevents future generations in that family from suffering its ravages. Lastly, some people have moral objections to destroying embryos simply because they carry a defect, but feel it is perfectly permissible, indeed potentially laudable, to treat the embryo by gene insertion to make it healthy.

In spite of the instant horror with which human genetic engineering is greeted, there may, after all, be substantial and real benefit. Moreover, if treatment of genetic disease by embryonic gene injection becomes feasible, why should we not go one step further and give our children a better start in life? There are many characteristics that people might think desirable, given the ability to implement genetic enhancement. They would almost certainly include beauty, intelligence, strength, aggression, prevention of ageing, and protection against disease. As it happens we already manipulate all of these in our children and young people in varying ways. In order to make our offspring more attractive we ensure they have orthodontic treatment for unsightly teeth, and we employ the best education possible, often on a selective basis at university and secondary school, to increase their mental capacity. Strength is encouraged by physical education and young sportsmen train vigorously to ensure they are as competitive as possible. Aggression is undoubtedly encouraged in young fighting men; the training for example in the marines or the infantry is largely designed to ensure this important fighting quality. We attempt to delay ageing in numerous ways, not least by taking hormone replacement therapy even before the menopause commences. Prevention of disease is a high priority in the welfare of children in nearly all societies, with the use of vaccination and vitamins on a regular basis.

What is ethically wrong with doing the same job more simply and permanently by inserting genes and creating 'designer babies'? The objections are manifold. Firstly, such treatment may produce an inherited elite, leading to increased tensions in society, such as racial prejudice. Because such treatments are likely to be the prerogative of the privileged, they might inevitably increase the poverty gap and class distinctions. There would also be a real risk of children not meeting the expectations of their parents after genetic manipulation – with consequent fragmentation of family and society. This is already a key issue with regard to screening the fetus during pregnancy. Western society has tended to screen for inherited disorders or handicap. When an abnormality is diagnosed, there is often said to be pressure to have a pregnancy terminated. Opponents of screening claim that it leads to the devaluation of handicapped people and possibly the same would be true in a society regarding genetic enhancement as being the desirable norm. A most important argument against genetic engineering to enhance our children's attributes is that we would be choosing on a contemporary and subjective basis what is regarded as being desirable. It may well be advantageous to be tall and blond in 1997, but in a different society in years to come, perhaps when the ideal is to be short and dark, such a blond descendant might be undesirable. He or she would have inherited a given characteristic that is not only unwelcome, but disadvantageous and permanent in all future generations.

Some philosophers have argued that allowing genetic enhancement to the potential offspring of interested couples is simply an extension of procreational liberties. This respects the autonomy of prospective parents. However, what is surely ethically unacceptable is to remove the autonomy of a future generation from choosing what it sees as in its best interests.

But, above all, the real ethical argument against introducing new genes into human embryos is the recognition that the effects of gene insertion are at present unpredictable. We could cause severe congenital handicap or disability. Mouse experiments have now been continued for about twenty years, but transgenic animals still frequently turn out to express unwanted or unpredictable traits than they do those desired by the scientists. There is no absolute guarantee that the DNA will be inserted into the correct part of the animal's chromosomes, the genome, or any certainty that other essential 'natural' genes may not be displaced or altered. This unpredictability is a cogent reason why transgenic technology is unlikely to be applied to humans in the near future. Leaving aside any considerations of morality or medical duty, no physician could afford to risk the medico-legal consequences. Not only can parents sue for a genetic mistake, but nowadays any deformed or disadvantaged child that results from such an experiment could successfully pursue a genetic engineering doctor through the law courts.

As it happens germ-line gene therapy to correct genetic disorders is a largely unnecessary technology. In order to treat an embryo by gene insertion, doctors would first need to know that it carried the defect needing treatment. With virtually all gene disorders only some of the embryos from a carrier couple will be affected. In order to confirm that an embryo is affected with a gene defect, preimplantation diagnosis would be needed. Given that access to embryos for gene manipulation is always likely to be by IVF and that several embryos will be available, it will be simpler – as well as safer – simply to transfer those embryos diagnosed free of the particular defect.

The notion of genetic manipulation of embryos for desired characteristics cannot, however, simply be dismissed on the basis of the availability of preimplantation diagnosis. Enhancement certainly would require the addition or altering of genes. Nevertheless, this too is extremely unlikely in the foreseeable future. Even if scientists in the long-distant future are able clearly to verify the complete safety of gene transfer in embryos, there would still be huge problems. Popular notions of genetic inheritance obfuscate the real issue. Beauty, intelligence, aggression and so on are not produced by the interaction of one or even a few genes. The genetic component of these traits is extremely complex but probably all the components need to be present. Take, as an example, the disease diabetes. Like height or strength, it has a strong genetic basis. Diabetes is a relatively simple disease, basically an intolerance to sugar which is caused by lack of insulin secretion from the pancreatic gland – unlike the trait of height or strength, a very simple but invariable mechanism. Yes diabetes is a multigenic disorder; we know of at least twenty separate genes on several different chromosomes which are likely to predispose to it. Consider how many different genetic interactions must have to take place for the creation of such 'desirable' qualities as beauty or intelligence. It is very unlikely that in the foreseeable future geneticists could unravel such complex knots and give us the DNA keys to enhancing our inheritance.

So much for what can be foreseen. Yet in the longer term, I think it very likely that germ-line gene therapy will be applied in humans because of its huge potential. The problem of its unpredictability will eventually be capable of solution by improved understanding of molecular biology and repeated, meticulous experimentation in various animal species. Paradoxically, the greatest value of germ-line gene therapy is likely to be in preventing multigenic diseases. Fifty per cent of us in the western world currently die of heart disease and the risk of developing coronary artery disease is highest in those people whose parents or grandparents suffered from it. Children of a diabetic parent have approximately between a 25 to 50 per cent chance of being diabetic themselves. Which of us carrying genes which predispose to these diseases, or diseases like Alzheimer's, would not prefer to see that our children have a reduced risk? By repairing or removing some of the genes in our germ cells which may induce these disorders, we may diminish these risks substantially.

The key could be to target sperm cells. Eventually, with merely a little local anaesthetic, it may be possible to inject protective genes directly into the testicle, where they could, with suitable chemical help, be incorporated into the developing sperms. With disorders which are multigenic, it will be possible to identify the best gene sequence to introduce to give at least partial protection. This would lower the risk of a susceptible family suffering, for example, from heart disease or diabetes. This technique would avoid all the complexities, expense and inconvenience of *in vitro* fertilization. It would be rather like vaccination but it would not only protect our children, but generations beyond. As well as reducing suffering, such an approach will have phenomenal economic advantages. The rising cost of health care – a major concern to all governments – could be controlled in future times. Such powerful technology will need the most stringent care if it is not to be abused, but its carefully regulated use seems eventually inevitable.

Cloning Humans

A clone is an individual, or a group of individuals, genetically indentical to another individual. Thus, identical twins – coming from one egg after spontaneous splitting following fertilization with a single sperm – are natural clones containing identical genes. Artificial cloning is not new. Humans have used plant clones for centuries, certainly since ancient Greece. MacIntosh apple trees, for example, are all a clone, having been produced from a single mutated plant, and all share identical genes. One of the first researchers to work seriously with animal clones was Dr Gurdon, who in 1968 published details of making a clone of a frog. He transplanted the nucleus from a tadpole's intestine into the egg of another frog. The egg had been previously prepared by destroying its own nucleus. Once the new nucleus took over function, the egg divided and grew into a mature frog – without any sperm being involved in fertilization. With the transfer of more nuclei from the same tadpole into more frogs' eggs, many identical frogs could be produced. Initially it was impossible to use the nucleus of a mature mammal for cloning, and until recently all experiments with cloning were done using a nucleus taken from an embryo. Amphibians such as frogs are a good deal easier to clone and it is only in the last fifteen years or so that scientists have found ways of making mammal clones of, for example, pigs and sheep. A major development occurred in 1997 when Dr Ian Wilmut and his colleagues from Edinburgh announced that it had finally been possible to use the nucleus taken from an adult animal and transfer it to an enucleated egg. Thus Dolly the sheep was born.

The most exciting aspect of Dr Wilmut's work is that the

nucleas from an adult cell is capable of being reprogrammed, being returned to 'infancy', where it can start to initiate the action of genes which only express in the earliest stages of development. Study of these and related phenomena will give us great insight into how cells are normally controlled and what goes wrong with this control – for example, in the genesis of cancers.

The ability to clone a mammal from an adult cell raised a spectre which still haunts the mind of the public. There was an immediate outcry when Wilmut and colleagues published their work. Quite serious and mature journalists argued that it was now merely a matter of time before some rich, elderly or powerful man, possibly an American – or, worse still, Saddam Hussein of Iraq – invested in this technology to reinvent himself. There was much talk of grieving people 'stealing' skin cells from some recently deceased relative and having the cells frozen for later cloning technology. Others suggested that misguided parents might want to clone children prematurely dead, possibly after being killed in an accident.

It is not easy to see a clear ethical objection to cloning a single individual on an isolated basis, not least because twins naturally occur already and their existence presents no obvious ethical problem. However, the thought of large numbers of clones from a single individual seems very different. But it is doubtful whether there could ever be any point in such an enterprise, unless it be to pander to some authoritarian dictator. It is, as we shall see, extremely unlikely that the authoritarian dictator would get much satisfaction from the fruits of his enterprise.

Much of the recent debate over cloning seems rather superficial and ill-judged. While there is a perfectly understandable objection to making 'carbon copies' of people, I do not believe that the technology behind cloning is nearly as threatening as it first appeared. Firstly, there is no chance at all that any individual created by nuclear transfer would be identical to the parent from whom the nucleus was taken. Although nearly all our DNA is held in the nucleus, a much smaller part of it is in the mitochondria – small organelles present in the cytoplasm of the egg. Consequently, the egg as well as the parent nucleus would contribute DNA to that individual. In genetic terms, that individual would be less similar to his or her parent than would identical twins be to each other. Moreover, we are undoubtedly a product of our nurture as much as our genetic nature. Even twins brought up together are entirely separate individuals with their own personality. In the case of a clone, the parent being an adult would inevitably have been brought up in entirely different circumstances from the cloned offspring. Who knows how a clone of a skin cell of an individual like Adolf Hitler might grow up given a warm and sustaining environment. He could be loving and gentle, and even contribute greatly to society.

There is a clear need to continue research into cloning. These technologies will have very significant applications for both human and animal well-being in the twenty-first century. There will be immense clinical value in being able to clone human tissues and organs, rather than whole people. Over the next decade or so there will be much research into the control of how cells differentiate. The embryo starts with just a few cells, each of which as we have seen, is totipotential – that is, capable of generating into a complete organism. The exciting challenge will be to manipulate embryonic cells so that they grow into just skin, or just muscle – or whatever tissue is needed. At present, when an individual contracts leukaemia there is a desperate search to find a compatible bone-marrow donor. Even if one is found, and a bone-marrow graft is successful, there follows a life of immune suppression with unpleasant and expensive drug therapy to ensure that the vital graft is not rejected. In the very far future, leukaemia could potentially be treated by taking the nucleus from one of the cells of the patient and then growing new bone-marrow cells which would be immunologically compatible; consequently there would be no rejection and no need to take immune suppressive drugs afterwards. This kind of approach, with modification, could be used for skin cells to provide skin banks for burns victims and other patients undergoing plastic surgery for various disfiguring lesions. It could be used to produce nervous tissue cells for transplantation into patients suffering from Parkinson's disease and other degenerative conditions. We could provide banks of muscle cells to treat the muscles of children weakened by various incurable forms of muscular dystrophy.

Cloning would also be of potential value for men suffering from intractable infertility. Some men cannot produce sperm because their testes are depleted of spermatogonia. One solution might be to make a clone. For many married couples, this would be ethically much more justifiable than using donor semen from, say, an anonymous donor. But refinements of cloning could allow for new cells which would effectively replace sperm. Sperm and eggs are different from all other cells in the body in having one copy of each of the normally paired chromosomes. This state prepares them for fertilization, so that the egg, when it develops, has paired chromosomes with one copy from the father and one from the mother. In time it may be possible to remove the nucleus of an adult cell having manipulated it to lose one set of chromosomes. This nucleus could then be injected into a normal egg which contains its own nucleus. This could then potentially develop like any fertilized embryo, with half its genes maternally derived and half paternally.

Another benefit of cloning would be in producing transgenic animals. Some scientists are presently claiming that the cloning of cattle might be used to produce whole herds with desirable traits – for example, very good milk yields, or excellent muscle for beef. This seems unlikely and I believe dangerous. Cloning a herd would result in loss of genetic diversity, which is a useful protective mechanism.

A genetically engineered herd which turned out to have little resistance to foot-and-mouth disease would be a liability, not an asset. Cloning might have much more use to produce transgenic animals, raised for medical purposes – such as organ transplantation. Recently there has been much interest in the use of transgenic pigs for transplantation. There is an acute shortage of spare hearts and other tissues, such as heart valves. Once a genetically engineered pig has been raised with the appropriate human genes, cloning would provide a convenient way of expanding the herd without loss of the unique genetic make-up needed for medical purposes. Cloning could also have important implications for the preservation and breeding of endangered species. Once initial clones had been obtained, they could subsequently be allowed to breed naturally so as to encourage normal genetic diversity.

Cloning, then, is unlikely to be used to produce multiple copies of similar or identical human beings. Although powerful technologies like this may cause considerable anxiety, there is little doubt that the benefits that could be derived from this research are really very significant. It should be perfectly possible to regulate these techniques in future, possibly by international agreement, so that they are used for the good, not to the detriment, of mankind.

Ectogenesis

In Aldous Huxley's novel *Brave New World*, human embryos were kept in vats, where they developed into fetuses. Birth was achieved by a process of decanting.

It is possible to keep rodent embryos outside the body for the early stages of development, and this has been applied experimentally in studies of placental development and in studies on toxicology. However, all mammalian embryos rapidly become too big to allow them to survive without a placenta supplying oxygen, nutriments and foodstuffs. There has never been the slightest success in developing an artificial placenta, which grows and adapts rapidly to the changing needs of the growing fetus. The development of an artificial placenta would be of great theoretical benefit in treating a number of congenital disorders. It would also help the understanding of the genesis of a number of diseases specific to pregnancy, for example toxaemia, and the causes of birth defects. However, given the complexity of placental function, and the extraordinary nature of its rapid growth and development, it seems most unlikely that scientists will be able to reproduce even a crude copy of this organ in the foreseeable future. For the time being, it seems, women will continue to give birth in the old-fashioned manner.

Contraception and the Population Explosion

Although humans are, relatively speaking, a very infertile species, they are extraordinarily prolific. Our intelligence and our ability to use tools means we protect ourselves from adverse environmental conditions. Consequently our fecundity now threatens the planet.

The issue of population growth and its control is certainly not new. As observed earlier, the ancient Egyptians used the expedient of infanticide to reduce the effects of the fertility of their Hebrew slaves. Much more recently, in the 1790s, Thomas Malthus proposed that Britain was heading for disaster. Continued expansion of the population from the then figure of around eight million meant that Britain would run out of food by 1900, when he miscalculated that there would be 112 million people to feed. As it happens, a number of unpredictable events such as improved hygiene and changing demography slowed population growth.

It is true that effective contraception is a very recent development and has already had profound social consequences in the improvement of our society. It has, above all, empowered women. Until this century, most women who were not prepared merely to accept menial or domestic activity had to choose between a career or marriage. Women who wanted a productive profession simply did not get married. Marie Curie, the pioneer physicist who isolated and purified radium, was one of the rare exceptions. This was because she was most unusually supported by her husband Pierre.

Unfortunately, world over-population is now a very genuine threat. In the years to 2010 we may face a population increase of around 15 to 25 per cent, around a thousand million people – more than the total population of India. Moreover most of these people will be born in the parts of the world least able to provide the resources needed for such an increase. There will be need for increased food, health services, employment, social security, sewage, maintenance of law and order, and education of course, on an unprecedented scale. Given that the world is already starting to see the effects of deforestation and global warming; that more pollution on a massive scale is almost inevitable; that social instability, both nationally and internationally is at risk; and, above all, that the population may continue to grow almost exponentially, these are very serious reproductive issues indeed. Under these circumstances, the manipulation of reproduction becomes one of the most important scientific issues of our time.

Contraceptive technology is clearly going to be one way of dealing with this burgeoning population. However, experience suggests that contraceptive technology is likely to be less important than improvements in health and education, because birth rates are invariably highest in the most underprivileged parts of the world. Unfortunately high birth rates contribute to social deprivation, so much of the developing world is caught in a vicious circle. Perhaps help to break this circle will come to a considerable degree from improved contraceptive methods.

The existing problems with current contraception are well known. Natural methods, such as breast feeding after pregnancy, and timing intercourse, are helpful but unpredictable and unreliable. The barrier methods need considerable commitment by couples, and are frequently felt to interfere with sex and its spontaneity. Moreover, many religious groups frown on their use. Hormone preparations avoid these problems but may have a number of side-effects which worry couples; they require considerable education and compliance from couples if taken orally. Intrauterine devices are associated with bleeding and menstrual cramping in a number of women. Sterilization has been increasingly used, but its irreversible nature and the fact that an operative manoeuvre is needed limit its applicability.

Over the next decade there will be continuing attempts to improve contraception. Various developments are likely. One area in which there has been more scientific investment is vaccination. It is likely to be possible to vaccinate women against certain proteins held in the envelope surrounding the egg, the *zona pellucida*. Because the *zona* is composed of protein and because there are several different proteins on the surface of the egg, it should be possible to derive a vaccine which would block the interaction of the sperm with the egg's surface. One of the problems with this approach, as with any immunological method to control conception, is that vaccination may result in irreversible contraception. However, it is likely that a vaccine with a short duration – perhaps of a year or so – could be developed. However, it is difficult to be precise about such factors, and people in poorer countries may not know whether they are still protected. Another concern about immunization against proteins in the *zona pellucida* is that the vaccine could result in an antibody reaction to all the eggs in the ovary, causing a premature menopause. This has been found by scientists working in Edinburgh to be a problem in experimental monkeys. It may be impossible to prevent this, in which case, of course, this technology will have little future.

Another approach, using vaccines, is to immunize a woman against sperm. Sperm antibodies are a well-known cause of infertility; women having them produce a reaction from their white cells which sees the sperm as 'foreign' and destroys them. Sperm antibodies immobilize any sperm which are not destroyed, and this is therefore potentially quite an effective method of birth control. The problem here is that sperm antibodies, once formed, are extremely difficult to get rid of. Vaccination may therefore be long-lasting, and perhaps the trick may be to find the best sperm protein to limit the length of immunity. So far it has proved surprisingly difficult in animal work to produce effective immunity against sperm at all.

One of the more promising approaches to immunization is to produce a vaccine against the hormones which are produced by the developing embryo. These hormones are proteins. The main pregnancy hormone is called human chorionic gonadotrophin (HCG). HCG is the hormone which can be detected in the bloodstream and urine about ten days after conception, and is the basis of the modern pregnancy test. It is a molecule composed effectively of two parts, a so-called alpha unit and a beta unit. If a large foreign protein is attached to the beta unit, it rejects. Preliminary vaccines have been developed and have been tested by the World Health Organization. Problems with the method include variability of immune response, variable duration, possible interference with other hormones in the body which have a rather similar structure, and the risk of causing miscarriages in later pregnancy. The last problem would be particularly serious in developing countries with poor economic and health resources.

One area which was thought promising, but in which too little research is currently proceeding, is the area of drugs which interfere with pituitary gland function in the brain. It is possible to block the production of FSH by giving so-called gonadotrophin-releasing hormone antagonists (GnRH). These drugs, which are used sometimes during IVF treatment and for the treatment of endometriosis, make a woman semi-menopausal by damping down ovarian activity. They can be given by depot injection once every month, or even every three months. As these drugs are particularly safe, the approach is rather attractive. The current problem is the temporary menopausal symptoms they cause, and also occasional irregular vaginal bleeding. It is possible in future that these drugs could be mixed with a small amount of oestrogen – as a kind of hormone replacement therapy – to provide a safe and symptom-free balance.

Envoi

It is a pretty foolish scientist who goes out on a limb and attempts to predict the future. In the 1830s members of the establishment predicted that it would be completely unsafe for humans to travel faster than thirty miles an hour. One is reminded of Leonardo da Vinci in the Court of Francesco Sforza of Milan in 1495. There he drew a heavier-than-air machine which was designed to fly, to the amusement and incredulity of the courtiers. Five hundred years later (a tiny space in man's history on this planet), a view of flight paths and schedules in and out of Milan International Airport might give those same courtiers pause for greater thought.

What is very clear, however, is that whatever happens in the future, human reproduction and its manipulation are going to pose some impressive and threatening challenges both scientifically and, more particularly, ethically. I do not share the horror of those people who are often illiterate about technology and who claim we have started to open a Pandora's box which threatens human welfare. It is fashionable in a society threatened by environmental deterioration and the visions of nuclear destruction to say that technology has gone too far. Yet we do not ban the modern methods of travel because of the greenhouse effect, nor the computer because of its possible use by belligerents. What we do is to regulate, modify and control the technology. Who is to decry Prometheus' gift of fire, simply because of the destruction of London in 1666, or the holocaust of Dresden in 1944? Whatever the future brings, we have to ensure that reproductive technology is used for human happiness and welfare, and to promote and protect human life and its dignity. That is a challenge which is worth taking and, I believe, achievable.

Glossary

Amino acid There are around twenty or so different amino acids. They are essentially the building blocks which combined together make the larger molecule, a protein. Proteins will all be different depending on which amino acids they contain.

Amniocentesis The investigation done at around twelve to seventeen weeks of pregnancy to assess the normality of the developing fetus. Fluid sucked from the sac surrounding the baby is analysed in the laboratory.

Antral stage Late stage of follicle (q.v.) development in the ovary when the egg starts to mature.

Apoptosis Cell death. Cells in all organs are continuously dying and being replaced. Apoptosis is the phenomenon of programmed or 'planned' cell death which is controlled by genetic influences. This is different from cell death from disease or injury.

Biopsy Removal of tissue or cells from the body or from an embryo for analysis.

Blastocyst Stage of embryonic development at around five days.

Chorionic tissue Cells from the developing placenta, sometimes biopsied for antenatal diagnosis of defects.

Chromosomes Structures on which the genes are placed. The human genome (q.v.) is essential to our genetic library, and each chromosome is a volume in that library. In humans there are twenty-three pairs of chromosomes, one-half of the pair inherited from the father, one from the mother. In addition there are two sex chromosomes, two copies of the X giving femaleness, and an X and Y forming a male.

Dominant genetic defect A genetic defect which, when inherited from either parent, invariably expresses and causes the disease. Consequently only one copy of the genetic mutation is required, on either chromosome, for expression (see also Recessive genetic defect).

Ectogenesis Development of the baby artificially, outside the uterus.

Endometrium The uterine lining into which the developing embryo implants.

Follicle The cystic structure in which an egg develops in the ovary. Initially about a millimetre in diameter, it grows to around 20 millimetres before rupturing at the time of ovulation.

FSH Insertion of genes into cells to combat genetic defects, or diseases such as cancer with a genetic component.

Genome The total composition of DNA in the body. The genome is effectively the blueprint of body function. The human genome project is the attempt to analyse the complete sequence of DNA in the genome.

Germ cell Either sperm or egg, or precursor cells which go to make these cells.

Glutamine An essential amino acid, required for normal life.

HCG Human chorionic gonadotrophin, the pregnancy hormone produced by cells in the placental tissue in the developing embryo.

HFEA Human Fertilization and Embryology Authority. The government regulatory authority set up in the UK to monitor and police all aspects of human IVF and donor insemination.

Hyperstimulation A serious condition caused by overstimulation by the drugs given to induce ovulation. On very rare occasions it has even been known to be fatal.

In vitro Biological processes (eg. fertilization of an embryo) which are activated outside the living body.

In vivo Biological processes which occur within the living body.

Intracytoplasmic Injection of a single sperm into the substance of an egg to induce fertilization in cases of severe male infertility.

Laparoscopy The insertion of a telescope into the abdomen to inspect the pelvic organs.

Laparotomy Surgical exploration of the abdominal contents through an open incision.

Lesch-Nyhan A crippling genetic sex-linked disorder affecting boys. The syndrome causes severe metabolic disturbances including kidney failure, and is one cause of physical handicap.

Mitochondria Small organelles present in every cell which contain a small part of our DNA. This DNA is mainly concerned with the genes which are responsible for some key metabolic processes.

Multigenic disease A disease caused by several genes interacting, or predisposing to it.

Nucleus The command structure of the cell which contains the great majority of the DNA.

Parthenogenesis Production of an embryo without sperm, that is without the process of fertilization.

Phenotype The physical characteristics of any organism which are largely determined by its genes (or genotype) but also influenced by environmental factors. Thus a child suffering from limb deformities caused by thalidomide will have a normal genotype – with normal genes – but an abnormal phenotype.

Preimplantation The diagnosis of defects, usually genetic or chromosomal, in embryos before the embryo has implanted in the mother's uterus.

Recessive genetic defect A genetic defect which causes disease only if the gene responsible is present on both the maternal and paternal chromosome. People having the gene defect merely on one of the chromosomes are said to be carriers, that is they can pass the gene on but are not affected themselves.

Recombinant Genetically engineered, i.e. produced by altering the composition of the DNA.

Sex-linked genetic defect A genetic defect inherited on the X chromosome. In affected cases, only males express the disease, while females may be carriers.

Somatic cell Any cell in the body which is differentiated for a particular function – such as skin, liver, gut – and therefore not a germ cell (q.v.).

Spermatogonia Primitive or precursor cells which in turn give rise to sperm.

Transgenic animal Any animal which has had new 'foreign' genes incorporated into it during early development. Because its germ cells will also be transgenic its offspring will inherit the same genes.

Zona pellucida The glycoprotein 'shell' or coat that surrounds and protects the unfertilized egg, and subsequently the embryo during the first five days after fertilization. At around five days, the *zona pellucida* ruptures and the embryo hatches, ready for implantation.